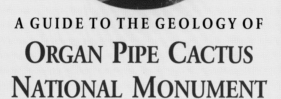

A GUIDE TO THE GEOLOGY OF
ORGAN PIPE CACTUS
NATIONAL MONUMENT

AND THE

PINACATE
BIOSPHERE RESERVE

ARIZONA GEOLOGICAL SURVEY
DOWN-TO-EARTH

JOHN V. BEZY
NATIONAL PARK SERVICE

JAMES T. GUTMANN
WESLEYAN UNIVERSITY

GORDON B. HAXEL
U.S. GEOLOGICAL SURVEY

Book design: Peter F. Corrao
Illustrations: Peter F. Corrao
Printed in the United States of America on recycled paper.

ACKNOWLEDGMENTS

The authors thank Carlos Castillo Sánchez, Guillermo Lara Góngora and Isabel
Granillo Duarte of the Reserva de la Biosfera El Pinacate y Gran Desierto de Altar for
their assistance with this publication. Thanks are also extended to Andrew W. Amann,
Jr., Marjorie Bengtson and Mitzi Frank of the National Park Service, Robert J. Miller
and Donald W. Peterson of the U.S. Geological Survey, and to David Bradbury of the
Los Alamos National Scientific Labroratory for editorial assistance. We are particularly grateful to Dr. Larry D. Fellows, Director of the Arizona Geological Survey, for his
editorial suggestions. Special thanks are due to Peter F. Corrao for the many hours he
devoted to designing this publication.

John C. Dohrenwend graciously granted permission to use his satellite images in
the design of this publication. Poster-sized satellite image maps (1:100,000) of the
Organ Pipe Cactus National Monument and the Reserva de la Biosfera El Pinacate
y Gran Desierto de Altar region can be ordered from Above and Beyond, P.O. Box
3021, Winter Park, Colorado 80482; phone: 970-726-4017; e-mail:
dohrenwend@rkymtnhi.com.

2000SP0300AZGS6832

CONTENTS

Acknowledgments   2

List of Figures   5

Introduction   6

Suggested Readings   62

Part 1:

Geologic History
of the Organ Pipe—
Pinacate Region   9

Part 2:

Organ Pipe
Cactus National
Monument   15

Ajo Mountain Drive

Feature 1:   Rock Varnish   *17*

Feature 2:   Beheaded Stream   *19*

Feature 3:   Rhyolite   *21*

Feature 4:   Talus Cones   *23*

Feature 5:   Childs Latite   *25*

Feature 6:   Desert Pavement   *27*

Feature 7:   Bajada and
             Pediment   *29*

Puerto Blanco Drive

Feature 8:   Alluvial Terraces   *31*

Feature 9:   Fault-controlled
             Springs   *33*

Feature 10: Mylonite   *35*

Feature 11: Granite   *37*

Part 3:

The Pinacate
Biosphere
Reserve   38

General Geology

Feature 12: Cinders   *43*

Feature 13: Aa and Pahoehoe
             Lava Flows   *45*

Feature 14: Cinder Cone   *47*

Feature 15: Playa   *49*

Feature 16: Tuff Cone   *51*

Feature 17: Maar Crater   *53*

Feature 18: Sierra Pinacate   *55*

Feature 19: Tuff Breccia   *57*

Feature 20: Dike   *59*

Feature 21: Eroded Pinacate
             Volcanics   *61*

LIST OF FIGURES

		Page
01	The Organ Pipe Cactus National Monument–Pinacate Biosphere Reserve region	10
02	Location of Features 1-11, Organ Pipe Cactus National Monument	11
03	Block diagram showing structure of northeastern part of Organ Pipe Cactus National Monument	12
04	Geologic time chart for the Organ Pipe Cactus National Monument–Pinacate Biosphere region	13
1-1	Rock varnish	16
2-1	Stream piracy	18
2-2	Beheaded stream in the valley of the south fork of Alamo Wash	19
3-1	The 26-14 million-year-old Ajo volcanic field and Organ Pipe Cactus National Monument	20
3-2	Western face of the northern Ajo Range	20
4-1	Talus cone along the western face of the northern Ajo Range	22
5-1	Typical outcrop of Childs Latite with characteristic large crystals of plagioclase	24
6-1	Desert pavement	26
7-1	A bajada along the southwestern margin of the Ajo Range	28
7-2	Cross-sectional sketch of bajada and pediment along southwestern margins of the Ajo Range	29
8-1	Alluvial terraces along the southeastern margins of the Cipriano Hills	30
8-2	Alluvial terrace surface of caliche-cemented basalt boulders and cobbles	31
9-1	Schematic geologic cross-section through southern Quitobaquito Hills and Quitobaquito Springs	32
9-2	Groundwater flow patterns that supply Quitobaquito Springs	32
10-1	Geologic sketch map of the Feature 10 area	34
10-2	Weathered surface of mylonite	34
10-3	System of deep-seated thrust faults within older rocks of Organ Pipe Cactus National Monument	34
11-1	Craggy outcrops of the Senita Basin granite, southeast of the Senita Basin picnic area	36
11-2	Geologic sketch map of the Sonoyta Mountains showing the extent of the Senita Basin granite	36
05	Location of Features 12-21, the eastern part of the Pinacate Biosphere Reserve	40
06	View of the Pinacate volcanic field from spacecraft	41
12-1	View looking northeast to the cinder cone (Mayo Cone) next to Feature 12	42
13-1	Aa flow near Tecolote Cone	44
13-2	Pahoehoe lava in the Pinacate	44
14-1	View looking north out through breach in Tecolote Cone to Mayo Cone	46
15-1	View looking northwest to the playa with Cerro Colorado in the foreground	48
16-1	Outcrop of tuff on the north rim of Cerro Colorado	50
17-1	Hypothetical cutaway of Crater Elegante	52
17-2	Aerial view of Crater Elegante from the southwest	52
18-1	View across Crater Elegante toward the peaks of Sierra Pinacate	54
19-1	Layers of tuff breccia that form rim beds of Crater Elegante	56
20-1	Dike of porphyritic basalt	58
21-1	East side of Sierra Pinacate	60

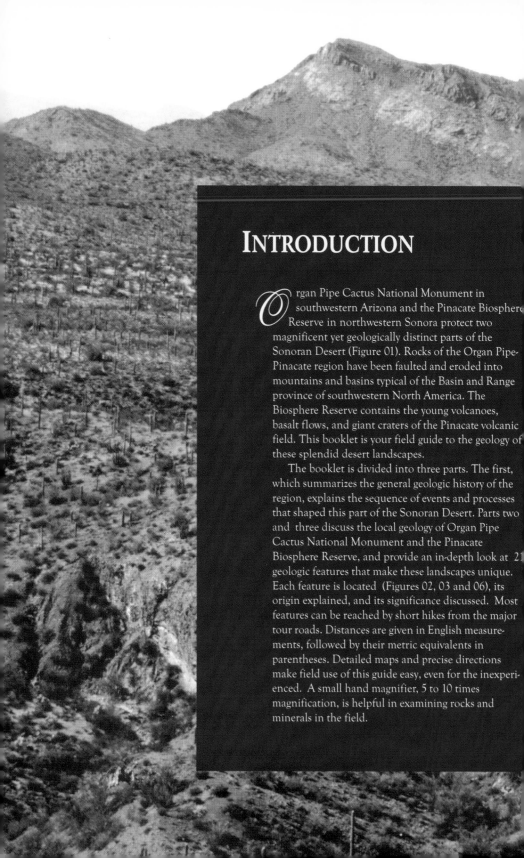

INTRODUCTION

*O*rgan Pipe Cactus National Monument in southwestern Arizona and the Pinacate Biosphere Reserve in northwestern Sonora protect two magnificent yet geologically distinct parts of the Sonoran Desert (Figure 01). Rocks of the Organ Pipe-Pinacate region have been faulted and eroded into mountains and basins typical of the Basin and Range province of southwestern North America. The Biosphere Reserve contains the young volcanoes, basalt flows, and giant craters of the Pinacate volcanic field. This booklet is your field guide to the geology of these splendid desert landscapes.

The booklet is divided into three parts. The first, which summarizes the general geologic history of the region, explains the sequence of events and processes that shaped this part of the Sonoran Desert. Parts two and three discuss the local geology of Organ Pipe Cactus National Monument and the Pinacate Biosphere Reserve, and provide an in-depth look at 21 geologic features that make these landscapes unique. Each feature is located (Figures 02, 03 and 06), its origin explained, and its significance discussed. Most features can be reached by short hikes from the major tour roads. Distances are given in English measurements, followed by their metric equivalents in parentheses. Detailed maps and precise directions make field use of this guide easy, even for the inexperienced. A small hand magnifier, 5 to 10 times magnification, is helpful in examining rocks and minerals in the field.

Desert Safety

The desert is a fragile but potentially dangerous place for the unprepared traveler. Before leaving home or camp, advise friends of your destination, route of travel, and time of return. Equip your vehicle with drinking water, maps, food, clothing, tools and signaling devices (mirror and flares). Roads are not paved and may be rough or sandy or have high centers, especially in the Pinacate. In case of mechanical failure stay with your vehicle and signal for help. Avoid unnecessary physical activity during the heat of the day.

Please protect the desert. Driving off approved roads is prohibited. Vehicle tracks are ugly scars that mar the landscape for years. Please do not litter or disturb plant and animal life. Bury human wastes at least 1 foot (.33 meters) underground. Bathing in or camping within one-half mile (800 meters) of natural water sources can deprive wildlife of water critical for survival. Use a gas camping stove rather than campfires. Weapons and traps are prohibited in these nature reserves. Leave the desert as undisturbed as its natural inhabitants have left it for you.

PART 1

GEOLOGIC HISTORY OF THE ORGAN PIPE–PINACATE REGION

*R*ocks exposed in the Organ Pipe and Pinacate reserves record many of the major geologic events that occurred in this region over the last 1.8 billion years. The record is not complete, however, because some extensive periods of geologic time are not represented in the rock record that remains today.

The oldest rocks are the 1.8 to 1.6 billion-year-old (early Proterozoic age; Figure 04) gneisses and granites that are exposed near the towns of Ajo and Sonoyta, in the Quitobaquito Hills, and near the Sierra Pinacate. These rocks, together with 1.4 billion-year-old granites of the Chico Shunie Hills southwest of Ajo and south of the Sierra Pinacate, represent early periods of continental crust formation and mountain building. The 550 to 330-million-year-old (Cambrian, Devonian and Mississippian) quartzite, schist and marble north of Growler Pass record a period of deposition of sediment in shallow seas. Granites and related gneiss near Sonoyta are the result of intrusion of granitic plutons 215 million years ago (Triassic age). The period from 180 to 145 million years ago (Jurassic time) was marked by several episodes of tectonic and volcanic activity and the intrusion of granite plutons, producing the low angle faults and the volcanic and granitic rocks of the Puerto Blanco and Agua Dulce Mountains and the Quitobaquito Hills. Metamorphosed conglomerates and breccias near Growler Pass represent sediments eroded from highland areas in late Jurassic to

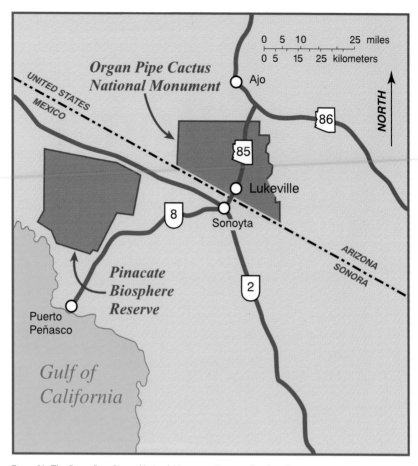

Figure 01. The Organ Pipe Cactus National Monument–Pinacate Biosphere Reserve region.

early Cretaceous time, approximately 150-140 million years ago.

The interval from late Cretaceous through early Tertiary time (80 to 40 million years ago) was marked by volcanism, intrusion of granitic rocks, and deformation of other older rocks. Rocks of this age include andesite, granodiorite and granite near Ajo (where they are associated with copper deposits), in the Quitobaquito Hills and Sonoyta Mountains and in the ranges near the town of Sonoyta. Movement on low angle faults occurred in several areas, notably the Puerto Blanco Mountains and Quitobaquito Hills.

The present landscape began to take shape 30 to 24 million years ago (middle Tertiary time) as the Earth's crust in the western part of North America was being stretched in an east-west direction. This stretching fractured the upper crust and its sedimentary and volcanic rocks into large, tilted blocks. Volcanism and the intrusion of granitic plutons resumed during the period 26 to 14 million years ago (middle to late Tertiary time), producing the andesite and rhyolite rocks of the Ajo volcanic field. During the period 15 to 10 million years ago, extensive

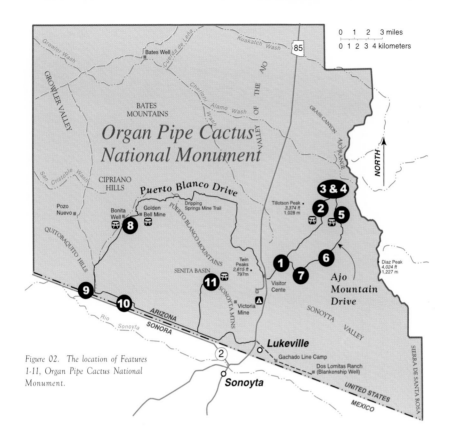

Figure 02. The location of Features 1-11, Organ Pipe Cactus National Monument.

faulting along steeply inclined surfaces uplifted the mountain ranges and lowered the structural valleys that form the modern Basin and Range topography of this region (Figure 05). From a few million years ago to several thousand years ago, eruptions of lava produced the cinder cones and basalt flows of the Pinacate volcanic field. Today, surface processes such as erosion by running water continue to shape the landscape.

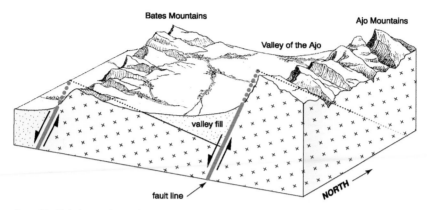

Figure 03. Block diagram showing the Basin
and Range structure of the northeastern part of
Organ Pipe Cactus National Monument.

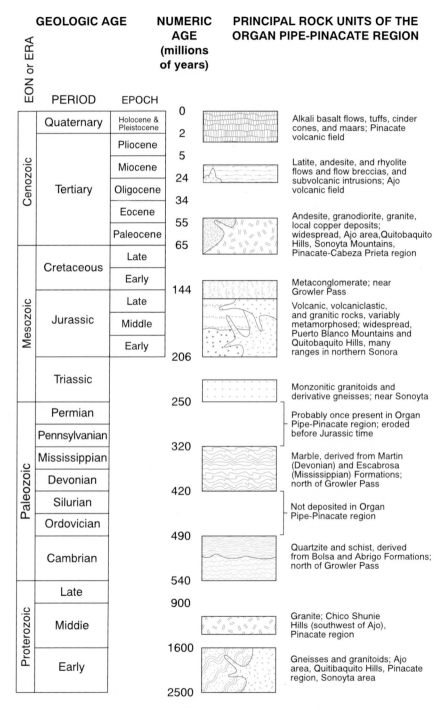

GEOLOGIC AGE			NUMERIC AGE (millions of years)	PRINCIPAL ROCK UNITS OF THE ORGAN PIPE-PINACATE REGION
EON or ERA	**PERIOD**	**EPOCH**		
Cenozoic	Quaternary	Holocene & Pleistocene	0 2	Alkali basalt flows, tuffs, cinder cones, and maars; Pinacate volcanic field
Cenozoic	Tertiary	Pliocene	5	Latite, andesite, and rhyolite flows and flow breccias, and subvolcanic intrusions; Ajo volcanic field
Cenozoic	Tertiary	Miocene	24	Latite, andesite, and rhyolite flows and flow breccias, and subvolcanic intrusions; Ajo volcanic field
Cenozoic	Tertiary	Oligocene	34	Latite, andesite, and rhyolite flows and flow breccias, and subvolcanic intrusions; Ajo volcanic field
Cenozoic	Tertiary	Eocene	55	Andesite, granodiorite, granite, local copper deposits; widespread, Ajo area, Quitobaquito Hills, Sonoyta Mountains, Pinacate-Cabeza Prieta region
Cenozoic	Tertiary	Paleocene	65	Andesite, granodiorite, granite, local copper deposits; widespread, Ajo area, Quitobaquito Hills, Sonoyta Mountains, Pinacate-Cabeza Prieta region
Mesozoic	Cretaceous	Late		
Mesozoic	Cretaceous	Early	144	Metaconglomerate; near Growler Pass
Mesozoic	Jurassic	Late		Volcanic, volcaniclastic, and granitic rocks, variably metamorphosed; widespread, Puerto Blanco Mountains and Quitobaquito Hills, many ranges in northern Sonora
Mesozoic	Jurassic	Middle		Volcanic, volcaniclastic, and granitic rocks, variably metamorphosed; widespread, Puerto Blanco Mountains and Quitobaquito Hills, many ranges in northern Sonora
Mesozoic	Jurassic	Early	206	Volcanic, volcaniclastic, and granitic rocks, variably metamorphosed; widespread, Puerto Blanco Mountains and Quitobaquito Hills, many ranges in northern Sonora
Mesozoic	Triassic		250	Monzonitic granitoids and derivative gneisses; near Sonoyta
Paleozoic	Permian			Probably once present in Organ Pipe-Pinacate region; eroded before Jurassic time
Paleozoic	Pennsylvanian		320	Probably once present in Organ Pipe-Pinacate region; eroded before Jurassic time
Paleozoic	Mississippian			Marble, derived from Martin (Devonian) and Escabrosa (Mississippian) Formations; north of Growler Pass
Paleozoic	Devonian		420	Marble, derived from Martin (Devonian) and Escabrosa (Mississippian) Formations; north of Growler Pass
Paleozoic	Silurian			Not deposited in Organ Pipe-Pinacate region
Paleozoic	Ordovician		490	Not deposited in Organ Pipe-Pinacate region
Paleozoic	Cambrian		540	Quartzite and schist, derived from Bolsa and Abrigo Formations; north of Growler Pass
Proterozoic	Late		900	
Proterozoic	Middie		1600	Granite; Chico Shunie Hills (southwest of Ajo), Pinacate region
Proterozoic	Early		2500	Gneisses and granitoids; Ajo area, Quitibaquito Hills, Pinacate region, Sonoyta area

Figure 04. Geologic time chart for the Organ Pipe Cactus National Monument–Pinacate Biosphere region.

PART 2

ORGAN PIPE CACTUS NATIONAL MONUMENT

*T*he 11 features described below represent some of the highlights of Organ Pipe geology. The first seven features are located along the most frequently visited scenic road in the monument, Ajo Mountain Drive (Figure 02). Features 8 through 11 are reached via Puerto Blanco Drive and the spur road to Senita Basin (Figure 02).

Features 3 and 5 introduce the reader to the colorful young volcanic rocks that dominate the landscape of Organ Pipe, whereas Features 9, 10, and 11 describe aspects of the older rocks of the monument. The remaining six features are the result of surface processes that shape this desert landscape today.

Ajo Mountain Drive

To begin the 21-mile (34-kilometer) Ajo Mountain Drive (Figure 02) exit the Monument Visitor Center parking lot and drive east across Highway 85. The loop drive takes approximately two hours without stops. Please remember to set your odometer at the start of the drive.

Puerto Blanco Drive

To begin the 53-mile (85-kilometer) Puerto Blanco Drive (Figure 02) exit the Monument Visitor Center parking lot to the west and follow the signs to the road. This trip takes approximately four hours without stops. Please remember to set you odometer at the start of the drive.

Figure 1-1. Rock varnish (coin and knife for scale).

GETTING THERE

Drive to the pullout at mile 1.75 on Ajo Mountain Drive; walk to the rock
cliff located approximately 100 yards (100 meters) north of the road.

ROCK VARNISH

1

he brown and black blotches (Figure 1-1) on these buff-colored rock surfaces are rock varnish. This mineral patina gives the desert landscape its tan to dark brown cast. Rock varnish develops best on weathering- and erosion-resistant rocks that have moderately rough surfaces. Sandstone, basalt, and many metamorphic rocks are commonly well varnished, whereas siltstones and shales disintegrate too rapidly to retain such a coating. Desert pavement (see Feature 6), that pebble- and cobble-covered ground surface so abundant in arid lands, commonly glistens because it is coated with this natural varnish.

Rock varnish is a thin coating (typically less than a hundredth of an inch) of clay minerals, such as illite, montmorillonite, and kaolinite, stained by heavy concentrations of iron and manganese oxides. The clay particles settle as dust from the atmosphere. Manganese, also derived from windborne dust and rain, produces a black to dark brown coloration on surfaces exposed to air. Micro-colonies of lichens and bacteria gain energy by oxidizing the manganese. They anchor themselves to rock surfaces with the clay particles, which provide protection against extremes in temperature and humidity. In the process, the manganese becomes firmly attached to and darkens the clay. Each time the rock surface is wetted by rain, more manganese and clay are brought in to sustain the slowly growing colony. Micro-colonies thrive where the rock acidity is neutral and the surface is so nutrient poor that competing colonies of lichens and mosses cannot survive.

Although older surfaces have darker colorations, desert varnish cannot be used as an accurate age-dating tool. The rate at which rock varnish forms is not constant and is affected by many variables, such as climate, wind abrasion, microclimate, biological competition, and abundance of manganese. Some researchers believe that the clay and manganese content of rock varnish reflects past climatic conditions. Because some varnished surfaces may be as much as 1 million years old, they could reveal valuable information about climatic change over a long period of time.

Well-varnished surfaces have a luster that causes entire hillsides to glisten in the intense desert sunlight. This mineral coating gives the landscape its warm tones of brown and ebony, often masking brilliantly colored bedrock below. Although all of the Earth's deserts have varnished rocks, varnish in the Southwest provokes even greater interest because of its archaeological importance. In innumerable locations prehistoric Indians pecked petroglyphs (rock writings) through the mineral skin to fresh rock below. Today these symbols are being revarnished as the process continues.

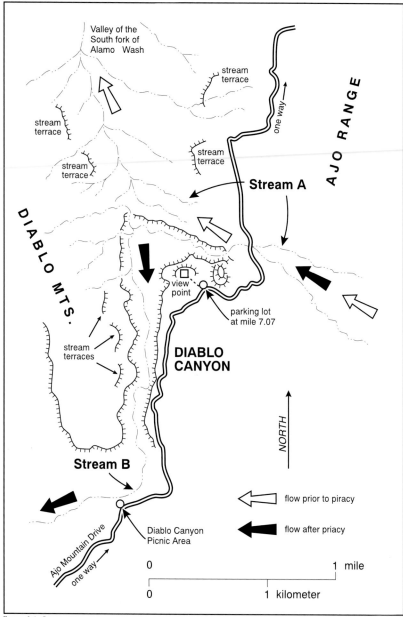

Figure 2-1. Stream piracy.

GETTING THERE

Drive to the small parking lot at mile 7.01 on Ajo Mountain Drive; walk
300 yards (274 meters) to the to the top of the low peak on the left
(north) side of the road for a full view of this feature.

FEATURE

BEHEADED STREAM

2

*S*treams are constantly enlarging their drainage areas, even in this arid environment. Over the course of repeated flash floods, rapidly eroding water courses can extend their headwaters across drainage divides and intercept the flow of less active streams. This process, called **stream piracy**, has occurred here at Canyon Diablo.

At one time stream A (Figure 2-1) drained to the northwest through the broad, gently sloping valley of the South Fork of Alamo Wash. Erosion was progressing more rapidly, however, along stream B. This stream was cutting along an easily eroded zone of rock shattered by a fault, and its steeper course moved flood waters at greater speeds than those of stream A. Expanding by headward erosion, stream B cut Canyon Diablo along this fault zone and eventually captured and diverted the flow of stream A to the southwest. From this vantage point, it is clear that this part of the valley of the South Fork of Alamo Wash (Figure 2-2, A) is abruptly truncated or **beheaded**—cut off from its former headwaters—by the deeper and younger Canyon Diablo (B). **Stream terraces** (see Feature 8), high remnants of former valley floors, can be observed along the walls of both valleys. Note that the surfaces of these terraces slope in the direction of flow of the stream that cut them. Stream piracy is common in mountains where hard and soft rock layers are exposed to erosion. In such terrain abandoned valleys and dry gorges, called wind gaps, are familiar evidence of shifting divides and diverted watercourses.

Figure 2-2. The beheaded stream in the valley of the South Fork of Alamo Wash.

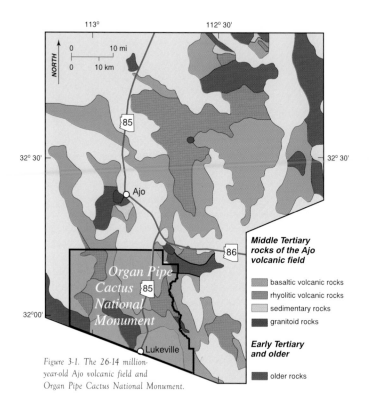

Figure 3-1. The 26-14 million-
year-old Ajo volcanic field and
Organ Pipe Cactus National Monument.

**Middle Tertiary
rocks of the Ajo
volcanic field**

- basaltic volcanic rocks
- rhyolitic volcanic rocks
- sedimentary rocks
- granitoid rocks

**Early Tertiary
and older**

- older rocks

Figure 3-2. The western face of the northern Ajo Range.

GETTING THERE

Drive to the turnout at mile 10.6 on Ajo Mountain Drive. This feature
is in the west face of the northern Ajo Range.

RHYOLITE, A VOLCANIC ROCK

3

*O*rgan Pipe Cactus National Monument includes part of an ancient volcanic field more than 1,900 square miles (4,921 square kilometers) in extent (Figure 3-1). Some of the most important and distinctive rocks of this great volcanic field are magnificently displayed here and at Feature 5, along the steep western front of the northern Ajo Range.

Geologists recognize two broad classes of igneous rocks—volcanic and plutonic. Volcanic rocks form by the rapid cooling of molten rock (magma) that has poured onto the surface as lava. Owing to rapid cooling, the crystals that make up these rocks are generally small. Plutonic rocks, on the other hand, crystallize from magma deeper within the earth and cool slowly. This allows the constituent crystals to grow to larger sizes than those of volcanic rocks.

The mountain in front of you (Figure 3-2) is composed chiefly of a volcanic rock called **rhyolite**. This rock formed from short, thick, bulbous lava flows. The ease with which such molten rock can flow is called viscosity. The viscosity of lava is determined by its chemical composition. All igneous rocks are composed of the same 11 essential chemical elements, but in widely varying proportions. Lavas that are low in silicon, such as the basaltic volcanic rocks of Hawaii, have low viscosity—that is, they are highly fluid (see Figure 13-1). The rhyolites of the northern Ajo Range, however, are relatively silicon-rich, substantially more viscous, and traveled only a few miles (kilometers) from their eruptive vents.

The massive, darker rocks in this mountain face are rhyolite flows. These flows were produced by relatively quiet, non-explosive eruptions. Many flows are separated by thin, light-colored layers of **tuff**—a rock composed of rhyolitic volcanic ash and small rock fragments ejected during shorter, explosive eruptions.

The Ajo Range is a great stack of volcanic flows and tuffs that were erupted in a brief but intense burst of volcanic activity 18 to 16 million years ago. Geologic mapping shows that this sequence is more than 2000 feet (610 meters) thick. Most individual flow and tuff layers cannot be followed laterally for more than a few hundred yards (meters). Some layers thin and disappear; others are bent, contorted, or abruptly truncated. Much of this complexity was caused by the high viscosity of the rhyolite flows.

Deformation along the bottoms, tops, and sides of the partially crystallized, sluggish lava flows produced **breccia**—a rock composed of large, broken, angular fragments. Pervasive sheer within the sticky lava produced thin convoluted layering called **flow banding**.

Figure 4-1. Talus cone (TC) along the western face of the northern Ajo Range.

GETTING THERE

This is the same location as Feature 3. The talus cones are located on the lower slopes of this western face of the Ajo Mountains.

TALUS CONES

4

*T*alus cones, those steep, triangular piles of rock rubble shown in Figure 4.1, are common landforms in high mountains and deserts. Wedging by ice and plant roots, decomposition and other types of weathering processes loosen rock fragments in the cliff. When dislodged, these chunks fall to the slope below the cliff and break into angular pieces that slide and tumble down the cone built by previous rock falls. Freshly fallen rock lacks the mineral coating called rock varnish (Feature 1) and contrasts sharply with the dark color of older debris.

Talus cones are the products of weathering followed by rock movement due to gravity. Both processes are important in reducing highlands and preparing the decomposed rock for removal by erosion.

These accumulations of fallen rock collect slowly and at variable rates. During extended periods of climatic cooling, when there is an increased number of daily freeze-thaw cycles, the rate of rock fall and accumulation is greater. Talus cones at Organ Pipe Cactus National Monument may be relict features produced during cooler and wetter periods of the Ice Ages (1,500,000 to 10,000 years ago). Many slopes in other parts of the monument were once covered with a nearly continuous apron of talus that is now being removed by erosion. Such talus deposits here show little evidence of recent rock falls and commonly contain boulders encrusted with lichens.

Talus slopes can be very unstable. Slope angles normally have reached their upper limit (angle of repose). Rockslides can be triggered when one attempts to climb the cones. Caution!

Figure 5-1. Typical outcrop of Childs Latite with characteristic large crystals of plagioclase.

GETTING THERE

Follow Ajo Mountain Drive to the Estes Canyon-Bull Pasture parking area and trailhead. From the trailhead walk about 40 yards (meters) east along the trail to the arroyo; turn right and walk about 60 yards (meters) south along the arroyo to the dark, reddish-brown, bouldery outcrop on the left (east) side of this drainage.

CHILDS LATITE, A SPECIAL VOLCANIC ROCK

5

ruption of the rhyolite lavas in the northern Ajo Range (Feature 3) was preceded by a slightly earlier volcanic episode that produced the distinctive rock you see here. This volcanic unit is called the **Childs Latite**. **Latite** is a type of volcanic rock; the name "Childs" comes from Childs Mountain near Ajo, where this rock unit was first described by geologists. Lava flows of Childs Latite, and related shallow intrusions, are exposed in more than a dozen mountain ranges in southwestern Arizona and northwestern Sonora.

The texture of volcanic rocks reveals a great deal about how they cooled. Throughout its wide distribution, the Childs Latite is characterized by large 0.5–1.5 inch (1–4 centimeters) crystals (Figure 5-1), called **phenocrysts**, enclosed in a matrix of much smaller crystals. The phenocrysts are plagioclase, one of the most common rock-forming minerals. These phenocrysts cooled and crystallized slowly in a subsurface magma chamber. Before crystallization was complete, this mixture of crystals and magma was erupted onto the Earth's surface as lava. Here the remaining liquid cooled rapidly to form the matrix of smaller crystals that surround the phenocrysts.

Pick up a piece of Childs Latite and closely examine the plagioclase crystals on a weathered surface. Try rotating the specimen into various orientations with respect to the sunlight. You will see that many of the phenocrysts contain subtle concentric bands or layers. These concentric zones are analogous to growth rings in trees. As tree rings record variations in growing conditions, so zoning in the phenocrysts reflects changing chemical conditions within the magma chamber. Such changes result from sudden cooling, loss of a batch of magma through eruption, sudden escape of gases dissolved in the magma, contamination of the magma by assimilation of foreign rock or mixing with another magma, or replenishment of the chamber by a fresh supply of magma.

The Childs Latite is an example of an alkaline igneous rock. Relatively uncommon, they are characterized by unusually high contents of alkali and alkali-earth elements: sodium, potassium, rubidium, strontium, and barium.

Geologic mapping of the Childs Latite reveals an interesting problem — the volcanic flows evidently were notably more fluid than might be expected for a phenocryst-rich magma of this composition. Perhaps its viscosity was lowered by an unusually high content of water or other volatile constituents.

Figure 6-1. Desert pavement.

GETTING THERE
Drive to the turnout at mile 13.27 on Ajo Mountain Drive.

DESERT PAVEMENT

6

his flat, cobble- and pebble-covered surface (Figure 6-1) is called **desert pavement**. It protects the finer material underneath from erosion by wind and water, especially when the surface cobbles and pebbles are cemented with mineral salts such as calcium carbonate.

Stone pavements such as this are produced by the wetting and drying of the desert surface. When certain clay minerals (smectites) in desert soils absorb rain water they expand to many times their dry volume. This swelling gradually moves the pebbles and cobbles within the soil to the surface. Over time, heaving from these repeated cycles of expansion and contraction produces a tightly fitted mosaic of stones on the desert surface. Removal of fine rock particles by strong winds may also play a role in concentrating the cobbles and pebbles in some pavements. Note that drying and contraction has produced cracks in the soil, but these quickly fill with sand and silt moved by wind and running water. Desert pavement, common to deserts everywhere, covers many thousands of square miles of the Earth's surface.

Figure 7-1. A bajada along the southwestern margin of the Ajo Range.

GETTING THERE

Drive to the turnout at mile 16.95 on Ajo Mountain Drive and climb
any of the nearby hills for the best view of these features.

BAJADA AND PEDIMENT

7

The broad plain (Figure 7-1) that sweeps southward from the base of the Ajo Range to the center of the Sonoyta Valley is a **bajada**. This graceful, sloping surface appears to be uniform in material, but consists of two very different sections: pediment and bajada. Both are hallmark landforms of western America's Basin and Range country.

The **pediment** is a planed-off bedrock platform (see Figure 7.2), an erosional feature carved by running water along the foot of the mountain front. Although the pediment at this point is covered by a thin veneer of sand and gravel, it is exposed in the walls of shallow drainages near the road. As wet-weather streams emerge from their confining canyon walls along the front of the Ajo Range, they assume a "side-winding " pattern, swinging laterally across the surface of the adjacent basin. In time, these streams level off the bedrock by cutting away any projecting rock hills, giving the pediment a smooth, graded slope. Sheetwash from thunderstorms keeps the pediment surface covered with a thin veneer of gravel.

Downslope, the pediment merges almost imperceptibly with the bajada. The latter is a thick sheet of gravel, sand, and silt that has been flushed from the canyons of the Ajo Range during flash floods. This rock debris first accumulates in fan-shaped deposits, called alluvial fans, at the mouths of these canyons. But as erosion wears back the mountain front the fans enlarge and eventually overlap, forming the bajada – that continuous apron of alluvial material that here fills the Sonoyta Valley .

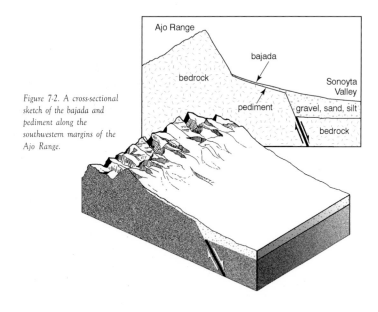

Figure 7-2. A cross-sectional sketch of the bajada and pediment along the southwestern margins of the Ajo Range.

29

Figure 8-1. Alluvial terraces (AT) along the southeastern margins of the Cipriano Hills.

GETTING THERE

Follow Puerto Blanco Drive to the Bonita Well Picnic Area. The
terraces are cut into alluvial fill along the southeastern flank of the
Cipriano Hills.

ALLUVIAL TERRACES

8

he flat, step-like land surfaces between the Bonita Well Picnic Area and the Cipriano Hills are **alluvial terraces** (Figure 8-1). As high remnants of earlier valley floors, their presence here is evidence of dramatic changes in the flow of Aguajita Wash through this valley over the past several thousand years.

Cobbles, sand, and gravel washed down from the Cipriano Hills and the Bates and Puerto Blanco Mountains once filled this valley to the level of the highest terrace (AT in Figure 8-1). Then climatic change and (or) changes downstream in the flow of the Rio Sonoita steepened the course of Aguajita Wash, causing its faster moving waters to cut down through and sweep away much of the older alluvial fill.

The process of downcutting halted at least five times, allowing the stream to erode laterally and produce floodplains. The stair-like flight of cobble-capped terraces in front of you is all that remains of these old valley floors. These surfaces are particularly resistant to erosion because they are capped with a cobble and boulder mosaic that is tightly cemented by the white calcium carbonate layer shown at point B on Figure 8-2. Uplift of the land, changes in rainfall, and stream piracy (see Feature 2), working separately or together, can cause erosion and associated terracing in any climate. In the desert, where their surfaces are relatively unvegetated and clearly seen, terraces are impressive testimony to the erosional power of running water.

Figure 8-2. Alluvial terrace surface of caliche-cemented basalt boulders and cobbles.

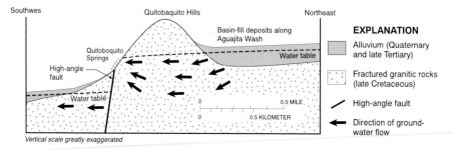

Figure 9-1. Schematic geologic cross-section through the southern Quitobaquito Hills and the Quitobaquito Springs, illustrating control of groundwater flow a by fault zone (after Carruth, 1996).

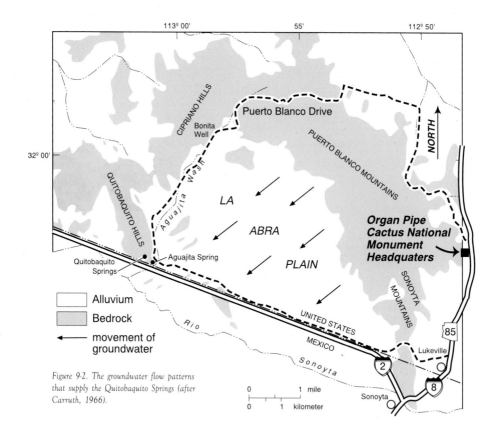

Figure 9-2. The groundwater flow patterns that supply the Quitobaquito Springs (after Carruth, 1966).

GETTING THERE

Follow Puerto Blanco Drive to the Quitobaquito Springs Road, turn right and drive to the parking lot.

FAULT-CONTROLLED SPRINGS

9

*Q*uitobaquito is an anomaly— a permanent, spring-fed pond amidst the aridity of the lower Sonoran Desert. Flow from Quitobaquito Springs varies seasonally and annually, but in recent years has averaged about 30 gallons (about 100 liters) per minute. The spring water is warm—77°F (25° C). The pond provides a home for a diverse assemblage of plants and birds and for the endangered desert pupfish. Three geological circumstances have caused localization of the Quitobaquito Springs: an impervious fault zone along the southwestern flank of the Quitobaquito Hills, highly fractured bedrock in the Quitobaquito Hills, and a large source of stored ground water beneath the La Abra Plain.

A large fault zone extends along the southwestern flank of the Quitobaquito Hills, the granitic hills just northeast of the pond. Rocks within the fault zone have been crushed into a fine-grained rock powder. This reduction in grain size speeded up chemical reaction with circulating ground water and caused the rock powder to be altered to clay minerals. The resulting fine-grained, clay-rich rock (called **fault gouge**) forms an impervious barrier that forces groundwater to rise to the surface as springs (Figure 9-1).

The ground water originates as rain falling on or adjacent to the La Abra Plain, a large desert basin bounded by the Sonoyta and Puerto Blanco Mountains and the Cipriano Hills and Quitobaquito Hills (Figure 9-2). This water percolates underground and flows southwestward through the gravel beneath the La Abra Plain. Upon encountering bedrock along the east side of the southern Quitobaquito Hills, most of the groundwater continues southward through the gravel beneath Aguajita Wash. A small portion of the water, however, passes southwestward through fractures in the granitic rocks of the Quitobaquito Hills and rises to the surface along the fault zone to feed the springs on the western side of the Hills. This flow pattern is driven by gravity because the La Abra Plain is 30 to 170 feet (9 to 52 meters) higher in elevation than Quitobaquito.

The powdery, light-colored fault gouge that controls the springs can be examined in numerous places along the edge of the Quitobaquito Hills, a few hundred yards (meters) north and northeast of the pond. The main spring is protected by a locked box, but a smaller spring in its natural condition can be examined about 50 yards (meters) due north of the east end of the parking lot.

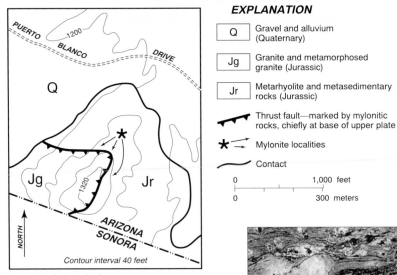

Q	Gravel and alluvium (Quaternary)
Jg	Granite and metamorphosed granite (Jurassic)
Jr	Metarhyolite and metasedimentary rocks (Jurassic)

Thrust fault—marked by mylonitic rocks, chiefly at base of upper plate

★⇁ Mylonite localities

Contact

```
0                    1,000  feet
0              300  meters
```

Contour interval 40 feet

Figure 10-1. Geologic sketch map of the Feature 10 area, showing topographic contours and exposures of the fault zone and mylonite.

Figure 10-2. Weathered surface of the mylonite. The larger, ovoid, partially flattened, and stretched crystals in this fine-grained matrix were shaped by intense ductile flow.

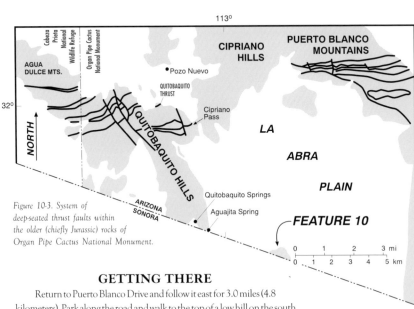

Figure 10-3. System of deep-seated thrust faults within the older (chiefly Jurassic) rocks of Organ Pipe Cactus National Monument.

GETTING THERE

Return to Puerto Blanco Drive and follow it east for 3.0 miles (4.8 kilometers). Park along the road and walk to the top of a low hill on the south (right) side of the road. If you are traveling west bound along the two-way portion of the Puerto Blanco Drive that parallels the international border, this feature is located 5.5 miles (8.8 kilometers) west of the junction with the Senita Basin road. The international border lies on the south side of this hill.

PLASTIC FLOW OF ROCKS: MYLONITE

10

*I*n everyday human experience, rocks are brittle: they break but do not flow. But within the Earth rocks can be either brittle or plastic. At shallow depths (less than about 3 miles; 5 kilometers), at low temperatures, and over short time periods (seconds to years) rocks generally are brittle. Earthquakes result from brittle fracture of rocks. At greater depths, at higher temperatures and over geologic time (millions of years), rocks generally behave plastically— they flow rather than fracture. Plastic flow does not require that the rocks be partially molten; rocks can flow while entirely in the solid state.

A familiar analogy is silly putty. If pulled rapidly silly putty snaps. This behavior is brittle failure. If pulled slowly it stretches like taffy. If a lump of silly putty is left on a table it flows as a viscous fluid over a period of an hour or so. Another useful analogy is road tar—stiff and brittle when cold but soft and plastic on a hot day.

The rocks that make up this hill record plastic flow in a deep-seated fault zone (Figure 10-1). The fault, well exposed near the top of the north-facing hillside, is inclined gently westward. The fault juxtaposes two different rock types: coarse-grained metamorphosed Jurassic granite above and fine-grained metamorphosed Jurassic volcanic and sedimentary rocks below. The fault itself is marked by a zone, several meters wide, of distinctive rock called **mylonite** (Figure 10-2).

This mylonite exhibits a matrix of small crystals of weaker minerals (especially quartz) that have undergone intense stretching and flattening by plastic flow. Embedded in this matrix are larger grains of stronger minerals. These larger grains were originally tabular or blocky in shape; their present ellipsoidal forms indicate that they have been partially flattened and stretched. Many of them have wispy "tails," where they have softened and begun to flow into the matrix. A few of the larger grains also show evidence of fracture. A particularly good example is visible in Figure 10-2, where a rounded feldspar crystal is offset along a small fracture.

Faults marked by mylonite are common within rocks formed deep in the Earth's crust. For example, the mylonite zone here is typical of an extensive system of faults that dominates the structure of the older rocks of Organ Pipe Cactus National Monument (Figure 10-3). These faults can also be examined in the area south of the northern Puerto Blanco Drive and southwest of Dripping Spring, or in the hills 1 to 2 miles (2 to 3 kilometers) southwest of Pozo Nuevo.

Figure 11-1. Craggy outcrops of the Senita Basin granite, southeast of the Senita Basin picnic area.

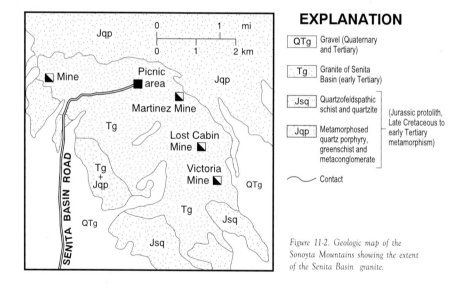

EXPLANATION

| QTg | Gravel (Quaternary and Tertiary) |

| Tg | Granite of Senita Basin (early Tertiary) |

| Jsq | Quartzofeldspathic schist and quartzite |

| Jqp | Metamorphosed quartz porphyry, greenschist and metaconglomerate |

(Jurassic protolith, Late Cretaceous to early Tertiary metamorphism)

~~ Contact

Figure 11-2. Geologic map of the Sonoyta Mountains showing the extent of the Senita Basin granite.

GETTING THERE

Continue east on Puerto Blanco Drive to the Senita Basin road. Turn left (north) and drive to the parking lot and picnic area at the end of the road. The Senita Basin road can also be reached from Highway 85 by driving 5 miles (8 kilometers) west along the two-way portion of the Puerto Blanco Drive that parallels the international border.

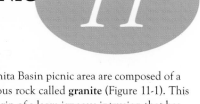

FEATURE

GRANITE, A PLUTONIC ROCK

he hills surrounding the Senita Basin picnic area are composed of a distinctive light-colored igneous rock called **granite** (Figure 11-1). This granite lies at the northern margin of a large igneous intrusion that has been weathered to form the low, rugged mountains that extend southward several kilometers from Senita Basin (Figure 11-2).

Geologists recognize two broad classes of igneous rocks—volcanic and plutonic. Volcanic rocks form by the rapid cooling of molten rock (magma) that has poured onto the surface as lava. Because of the rapid cooling rate, these rocks are typically composed of small crystals. Plutonic rocks, named for Pluto, ruler of the underworld in Roman mythology, crystallize deep within the Earth and cool slowly. This allows the constituent crystals to grow to larger sizes than those in volcanic rocks. The Senita Basin granite is an excellent example of a plutonic rock.

The Senita Basin granite is a plutonic rock that consists of approximately equal amounts of three minerals: quartz, alkali feldspar (rich in potassium and sodium), and plagioclase. To identify them, examine both weathered and fresh rock surfaces. The quartz is glassy-looking and clear to gray; plagioclase is grayish but not as glassy as quartz and bears fine striations on some grains; the alkali feldspar is generally blocky in shape and frequently slightly reddish or orangish in color. The granite also contains small amounts of several other minerals: shiny flakes of black mica (biotite) and white mica (muscovite) and, in some places, small garnet crystals a few millimeters in diameter.

The Senita Basin granite crystallized very slowly from magma about 60 million years ago at a depth of 6 to 9 miles (10 to 15 kilometers). Even though the pluton was entirely solid rock it retained a great deal of internal heat. This remnant heat provided the energy to drive a great thermal convection system as hot, water-rich fluids that contained minerals in solution circulated through fractures in the cooling pluton. The migrating fluids left evidence of their passage in the form of minerals deposited along fractures. In walking over the Senita Basin granite, you will see that many fracture surfaces are coated with veins or seams of shiny, finely crystalline muscovite. Veins of milky white quartz also are common.

A few of the larger quartz veins bear minerals containing lead, silver, copper, gold, or zinc. In the late 1800s and early 1900s (prior to establishment of the National Monument) small mines or prospects were developed in some of these metal-bearing quartz veins. Three old mines—Victoria, Lost Cabin, and Martinez (Figure 11-2)—are located along the margins of the pluton.

General Geology

*T*he Pinacate Biosphere Reserve comprises more than 580 square miles (1,500 square km) dotted with hundreds of small volcanoes and nine large craters that formed during explosive eruptions (Figures 06 and 07). Pinacate volcanic activity began a few million years ago and has continued to recent times. Some of the small volcanoes have been little modified by erosion and their lava flows look almost as if they cooled only yesterday. Indeed some could have erupted only a few thousand years ago. Pinacate volcanism probably is not extinct but only dormant.

Most of the volcanic rocks here are relatively rich in iron and magnesium compared with those at Organ Pipe Cactus National Monument. Iron and magnesium tend to form dark-colored minerals when the molten lava cools and crystallizes. Hence the resulting rock is dark gray in color and is known as **basalt**. Basalt is the same kind of lava that forms the Hawaiian Islands, Iceland, and the floors of the oceans. It is the most common type of volcanic rock on Earth.

The Pinacate is very near the Gulf of California, where two of Earth's great tectonic plates are separating from one another, just as they are in Iceland. However the origin of Pinacate magma (molten rock) may be quite different from that in Iceland and in the Gulf of California. Geologists are uncertain how the Pinacate came to be where it is. They know that basalt magma forms when a mass of hot, largely solid rock many tens of kilometers down in Earth's mantle rises slowly toward the surface. As this mass rises, the pressure on it decreases and it becomes partially molten. The resulting melt is forced by the weight of the overlying rocks to flow up through cracks to the surface of the Earth.

The rising basalt magma was relatively runny and contained bubbles of gas. As it approached the surface, these gas bubbles expanded and finally burst, causing the magma to spray into the air as frothy blobs of liquid. These blobs fell to the ground around the vent forming small volcanoes known as **cinder cones**. Some of the magma contained fewer gas bubbles and poured quietly out across

PART 3

the ground as lava flows. At other localities, rising Pinacate magma encountered large quantities of water under the ground and heated the water to produce steam explosions. These explosive eruptions formed large craters for which the Pinacate is known. You will see all of these features and more in the Pinacate Biosphere Reserve, which is a remarkable showcase of many products of basalt volcanism.

Entry to the Pinacate

Drive 32 miles (52 km) on Mexican Highway 8 from the bridge in Sonoyta toward Puerto Penasco (Figure 05). Take the first right turn after a bridge and register at the Estacion Biologica, the visitor center of the Pinacate Biosphere Reserve. If you wish to climb Mayo Cone (Feature 14) or to enter Crater Elegante (to Features 19 and 20), permission to do so should be obtained at this time. Follow the dirt road 6.6 miles (10.6 km) north to a fork leading off to the left and signed to Crater Elegante. Bear left at this fork. [The right fork leads 10.8 miles (17.4 km) north to Feature 16.] After taking the left fork, continue 8.5 miles (13.7 km) to a three-way intersection. Turn right (east) at the intersection for 2.9 miles (4.7 km). Veer left (north) at the next intersection and continue for 2.4 miles (3.9 km), where you should turn back sharp left on the road signed to Tecolote.

Especially in the event of wet roads, you may enter the Pinacate from the north by a shorter, rockier route. You will miss the Visitor Center and should register with and obtain necessary permission (see above) from Reserve authorities at the first opportunity. To enter from the north, turn south off Mexican Highway 2 onto the cinder road 32.2 miles (51.8 km) west of Sonoyta. Follow the road for 3.8 miles (6.1.km). Near the gate at the end of the road, turn right (west) on a rough dirt road and follow it 3.6 miles (5.8 km) to a fork. Take the right (west) fork to Tecolote.

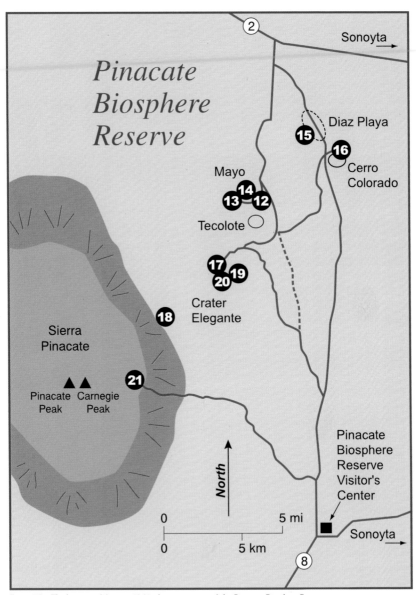

Figure 05. The location of features 12-21, the eastern part of the Pinacate Biosphere Reserve.

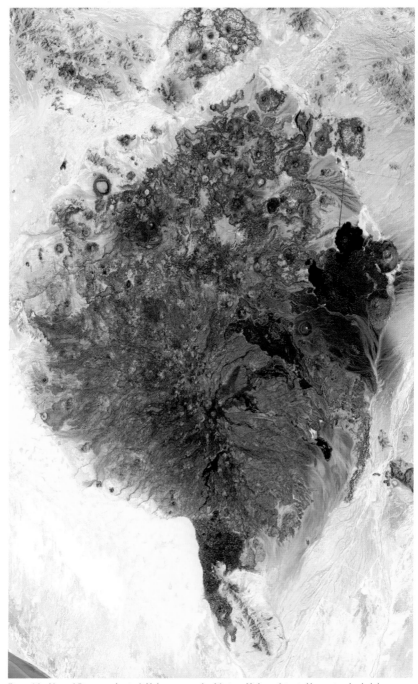

Figure 06. View of Pinacate volcanic field from spacecraft. Mexican Highway 2 is visable crossing the dark lavas near the northern end of the field.

Figure 12-1. *View of the cinder cone (Mayo Cone) next to Feature 12.*

GETTING THERE

Follow the Tecolote road 0.7 miles (1.1 km) to several old bulldozer pits
dug in the cinders on both sides of the road.

CINDERS

he surface of the ground in this area is made up of gray, frothy fragments of rock known as volcanic cinders. The pits show that these fragments have accumulated to a depth of at least several meters here. They originated from the vent of the nearby cinder cone, Mayo Cone. Note how light in weight the cinders are. The bubbles in them were originally steam bubbles that formed in the following way.

When the basalt magma was produced far down in the mantle, it contained small amounts of gas, chiefly steam, carbon dioxide and sulfur dioxide. These gases were dissolved in the magma just as carbon dioxide stays dissolved in soda pop until you open the bottle. As the magma approached the surface, the pressure on it decreased greatly and the gas formed bubbles that caused the magma to spray out of the ground, much like the soda pop would if you shook the bottle and then released the pressure inside by opening the lid. The resulting frothy blobs cooled and hardened as they flew through the air, becoming solid fragments known as cinders. The cinders accumulated around the vent to form a cinder cone.

Note the size of the cinder fragments here so that you can compare them to the size of those fragments you will see near the vent of a cinder cone (Figure 12-1; Feature 14). The largest fragments tend to accumulate near the vent, whereas the small and light-weight cinders are more readily carried off by air currents and are widely distributed around the erupting volcano.

Figure 13-1. Aa flow near Tecolote Cone.

Figure 13-2. Pahoehoe lava in the Pinacate. Note the ropy surface of this relatively runny basalt flow. Knife gives scale in the photograph.

GETTING THERE

Continue west for 0.2 mile (0.3 km) to Tecolote campground and park. Feature 13 is at the end of the rocky road 0.3 mile (0.5 km) west of the campground.

AA AND PAHOEHOE LAVA FLOWS

*T*hree basalt flows floor and surround the oval basin at the end of the road. The very rough-surfaced flows that surround the basin are most conspicuous. These are known as **aa** (pronounced *ah ah*) flows (Figure 13-1). The one on the south side of the basin came from Tecolote Cone, the large cinder cone nearly 1 mile (1.6 km) to the SSE. It is one of the youngest lava flows in the Pinacate. The one on the north side of the basin came from the cone very nearby to the NNE (Mayo Cone). The surface of contact between these two young aa flows is hard to find. Walk a short distance across the flow to the south and you will soon appreciate the rough character of its surface.

Basalt lava becomes more pasty and stiff as it cools, and, therefore, flows more slowly. The rough surfaces of aa flows develop as lava moves away from the vent, cools, and loses its dissolved gas. The surface of the lava becomes brittle when it has cooled and is commonly broken up by the motion of the still-hot lava beneath. Also, pasty clots of magma develop in the moving flow and work their way up to the surface. Thus, the surface becomes littered with jagged slabs and chunks of solid lava.

On the smooth floor of the basin, especially along its western margins, is the surface of another lava flow that is slightly older than the two aa flows. This surface is smooth and billowy and is typical of the type of lava flow known as **pahoehoe** (pronounced pa–hoy–hoy). Although they are rare here, pahoehoe flows sometimes have ropy wrinkles on their top surfaces (Figure 13-2). These wrinkles form as the hot, still-plastic skin of the flow is dragged by motion of the liquid beneath. They suggest that the magma was not stiff but may have flowed about like warm tar. Note the abundant gas bubbles in the pahoehoe lava here. Pahoehoe flows occur when basalt magma is still relatively rich in gas and very hot, perhaps 2,100°F (1,150° C).

Figure 14-1. View looking north out through the breach in Tecolote Cone to Mayo Cone. Pits in the foreground are explosion craters produced during the last eruptions at Tecolote. The small hills visible in the middle distance between these craters and Mayo Cone, especially on the left side of the photo, are large masses of cinders. These cinders formerly were part of Telecote and were carried away from it when a lava flow poured north from the cone.

GETTING THERE

Return to Tecolote campground and with permission from the Reserve, take the obvious trail that ascends gradually up the southwest side of Mayo Cone, directly to the north. Please stay on the trail, which winds up to the rim of the cone. When you descend, again stay on the trail and do NOT ski down the side of the Pinacate cone.

46

CINDER CONE

14

As you climb the cinder cone north of the campground, note that the cinders become much larger than those you saw at Feature 12. Masses of fresh magma more than 2.6 inches (66 mm) across are known as **volcanic bombs.** They are abundant at and near the rims of Pinacate cinder cones. From the highest point on the rim of this cone one can view the pinkish tan tuff cone named Cerro Colorado (Feature 16) off to the east and also of Tecolote, the large young cone about one mile south of the campground.

The main vent for the eruption that built Mayo Cone lay within its central crater. Note that this crater is open (breached) to the west. Most Pinacate cinder cones are breached. At some localities the breach was kept open by continuous flow of lava during cone-building eruptions. At many others, including this one and Tecolote, the breach was opened one or more times when lava welled up within the cone and broke out through one side of it as a lava flow. These flows moved large masses of cinders away from the breach. The lava that flowed west from Mayo carried much cinder, as did a large flow that moved out from the breach in the northwest wall of Tecolote (Figure 14-1). The rim of Mayo Cone affords a particularly good view of the terrain north of Tecolote.

Best seen along their eastern rims, both Mayo and Tecolote display scarps where masses of cinder slumped downward during eruptive activity. Slumping down into the vent region is especially obvious. The crater of Mayo contains several pits that were formed by explosions that occurred during the end stages of the cone's eruptive history. Similar pits are evident in the breach of Tecolote (Figure 14-1).

Tecolote is younger and larger than Mayo and its history is more complex. Slump scarps on top of Tecolote can be yellowish to reddish due to chemical alteration by sulfurous, acid vapor seeping through the cone. Some other features of Tecolote include enormous, dense bombs littering its surface on the south side and the presence of several small lava flows that leaked from the southeast wall of the cone near its base.

Although it appears very young, Mayo cone evidently is at least about 25 thousand years old. This age highlights the very slow rate of natural geomorphic change in this arid region and indicates that scars we make on the landscape with vehicles or on foot will be visible for a very long time. Please stay on the trail up and down any "young" cinder cone you climb!

Figure 15-1. View looking northwest over Cerro Colorado in the foreground to the playa (light-colored oval) in the upper half of the picture. The black area southeast of the playa is a very young lava flow.

GETTING THERE

From Tecolote campground, return 0.9 mile (1.4 km.) to the junction with the main park road and turn south (right). Proceed 2.4 miles (3.9 km) to a fork. Go straight (left fork) about 75 yards (68 meters) and then take another left fork (turn east). Drive 4.3 miles (6.9 km) to an intersection with a north-south road. Diaz Playa lies about 1.5 miles (2.5 km) north.

Caution: Do not drive across the playa after a rainstorm!

PLAYA

15

*D*iaz Playa, the pale tan expanse northwest of Cerro Colorado, is a dry lake bed (Figure 15-1). A south-flowing drainage was blocked by the eruption of Cerro Colorado tuff cone, resulting in the periodic ponding of water on the playa after heavy rain. Evaporation rates are high in this arid climate and rainwater that collects on this playa after storms soon disappears as it evaporates and soaks into the ground.

Diaz Playa is just one of hundreds of wet-weather lake basins in the deserts of North America. Sediment from several of these playas has yielded fossils and pollen that provide evidence of climatic conditions, vegetation and animal life during the last Ice Age. This playa has not been cored for such evidence, nor is its date of formation known.

The pale tan color of the playa is due to fine clay washed in from the north and northeast. Winds blowing across this and other sparsely vegetated playa surfaces pick up these clay particles and spread them as dust across broad regions, where they play a role in rock weathering and the formation of rock varnish.

Diaz Playa is named in memory of the first Spanish conquistador to explore this region, Melchior Diaz, who died east of here in 1541 after visiting the Colorado River.

Figure 16-1. Outcrop of tuff on the north rim of Cerro Colorado. Note the continuous layers of near-constant thickness. These bespeak deposition by direct fallout from eruption clouds.

GETTING THERE

From the intersection of the N-S and the E-W roads south of Diaz Playa
and just west of Cerro Colorado, drive east for about 0.4 mile (0.6 km)
to the north rim of the tan cone.

TUFF CONE

16

*Y*ou are standing at the rim of Cerro Colorado ("red hill"), a young volcano evident in the foreground of Figure 15-1. This cone is composed chiefly of volcanic ash (small bits of ejecta). A volcanic rock made of ash is called **tuff**. Cerro Colorado is a superb example of a tuff cone.

Look closely and you will see that the tuff consists of pieces of gravel and scattered fragments of basalt in a matrix of olive to orangish tan pellets about .05 to .2 inches (1-5 mm) across. When basalt magma rose to the surface, it encountered abundant water in the ground. The heat of the magma flashed the water into steam and the magma itself was blown into tiny fragments of tan ash. The pieces of gravel in the ash represent old stream gravel that underlies the volcano. The explosions tore loose and erupted both the gravel and a few chunks of basalt from an old lava flow visible on the southwest side of the floor of the crater.

Abundant water in the eruption clouds caused the accumulating ash to be relatively wet and sticky, especially in the upper parts of the cone. Here, layer on layer of ash fell to the ground from the repeated steam explosions (Figure 16-1). Eruptions occurred from several adjacent circular vents. The crater itself, with its somewhat "cloverleaf" outline, represents the coalescence of these vents as their walls periodically collapsed.

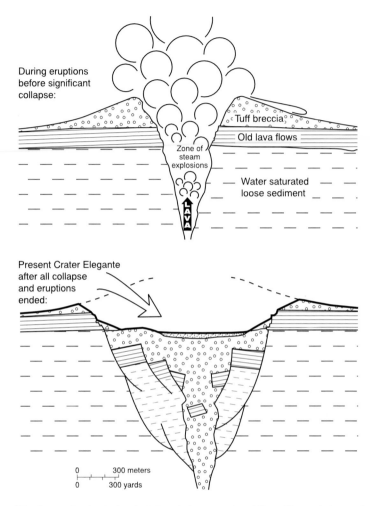

During eruptions
before significant
collapse:

Tuff breccia

Old lava flows

Zone of
steam
explosions

Water saturated
loose sediment

LAVA

Present Crater Elegante
after all collapse
and eruptions
ended:

| 0 | 300 meters |
| 0 | 300 yards |

Figure 17-1. Cross section showing hypothetical, vertical, cutaway view of Crater Elegante during eruptions (top) and as it is seen today following all eruptions and collapse to form the present crater (below). Note that the three major types of rock or sediment involved are labeled on the top diagram and have unique symbols. The same symbols are used to denote them on the lower diagram. A thin layer covering the tuff breccia on the floor of the present crater is shown with a dotted pattern on the lower diagram and represents recent sediment washed across the crater floor. Also shown on the lower diagram are accumulations of talus at the foot of the cliffs; the talus is denoted with a triangular symbol.

GETTING THERE

Return 4.7 miles (7.6 km) toward Tecolote and continue west (left) from the intersection on the road signed to Crater Elegante. Drive 2.9 miles (4.7 km) to a road intersection and continue west 1.6 miles (2.6 km) to a parking area at the end of the road. Walk up to the rim of the crater.

MAAR CRATER

17

rater Elegante (Figure 17-2) is 1 mile (1.6 km) in diameter and 800 feet (244 meters) deep. It formed about 150,000 years ago during catastrophic eruptions in which basalt magma interacted explosively with water in the sediment below the volcano. This flashed the water into steam. Exposed in its cliffs are numerous lava flows, most of which are unrelated in origin to the present crater. Part of the crater formed by explosive ejection of these older rocks from the vent. Most of the crater, however, originated by collapse of the volcano and the older lavas beneath it. These lavas rested on mud. As huge quantities of mud were ejected together with volcanic ash, this underlying support was removed. Masses of volcanic material crumbled into the vent and the volcanic pile finally subsided to form the present crater (figure 17-1). The bottom of the crater lies far below the level of the preexisting ground surface. Such a crater is known as a **maar**.

About 15,000 years ago, when the climate here was much cooler and wetter, a lake occupied the bottom of the crater. Sediment deposited along the shore of that lake forms a sort of "bathtub ring" that is especially evident on the west side of the crater as much as 200 feet (60 meters) above its floor (Figure 18-1).

As the volcano collapsed to form the crater, an old cinder cone was sliced nearly through its center. The interior structure of this cone is well displayed in Elegante's southeastern cliffs.

Figure 17-2. Aerial view of Crater Elegante from the southwest. Crater diameter is 1 mile (1.6 kilometers). An old cinder cone lies south of the crater. Part of another cinder cone can be seen here in the far crater wall. When you first come from the road to the rim of the crater, you are standing almost exactly across the center of the crater from where this picture was taken.

Figure 18-1. View across Crater Elegante toward the peaks of the Sierra Pinacate. Some very young lava flows in the middle distance appear as a thin band of black rocks just above the far rim of the crater. Sediment deposited in the lake that used to occupy the lower part of Crater Elegante is visible as a ring-like bench about 200 feet (60 meters) above the crater floor.

FEATURE

SIERRA PINACATE

18

orming the skyline southwest of Elegante is the Sierra Pinacate (Figure 18-1), a volcanic mountain that rises to an elevation of 3,904 feet (1,190 meters) above sea level. Jesuit missionaries, most prominently Padre Eusebio Francisco Kino, recognized the volcanic origin of this terrain as early as 1701.

Eruptions probably began in the Pinacate a few million years ago; geologists do not know exactly how long ago because the oldest volcanic rocks are buried and not accessible for study. Until about one million years ago, eruptions built a rounded, gently sloping pile of volcanic rocks known as a **shield volcano**. The character of the erupted lava changed gradually with time until eruptions produced a type of lava much poorer in iron and magnesium and lighter in color than basalt.

For about the last one million years, Pinacate volcanic activity has produced basalts from vents scattered over a broad region. These basalts covered the slopes of the old shield volcano (Santa Clara) and much surrounding terrain. Most of the volcanic rocks exposed in the Pinacate belong to this younger group of lavas.

Both of the prominent peaks atop the Sierra Pinacate are cinder cones. Pinacate Peak, on the west, is slightly higher than Carnegie Peak, which was named during the 1907 Carnegie Institution expedition to the Pinacate. Carnegie is a young cone. If the light is favorable, you may clearly see lava flows that ran down the side of the Sierra Pinacate from vents near Carnegie Peak.

Figure 19-1. Layers of tuff breccia (tuff containing large blocks of basalt) that form the rim beds of Crater Elegante.

GETTING THERE

The next two features are located part way down inside the crater. As of this writing, you need permission to enter Crater Elegante (see Entry to the Pinacate). The trail to Features 19 and 20 begins about 2,600 feet (800 meters) south around the crater rim. This trail leads about one quarter of the way down into the crater. It is steep and passes over rough terrain and near cliffs. It should be undertaken only by those in good physical condition and with sturdy hiking boots. Use it only at your own risk. Remember that you have to come back up. Accordingly, you may choose to skip Feature 20 and to view rocks like those at Feature 19 about 30 to 40 yards (meters) into the crater from the point where you first reached its rim.

TUFF BRECCIA

19

*A*s you enter the crater you will see outcrops of tan, layered, crumbly rock. The eruptions that led to Crater Elegante blew out large quantities of tan silt and clay as well as frothy pellets of ash representing the magma whose heat powered the eruption. The explosions also tore loose fragments of old, solid basalt from the walls of the vent. These fragments range in size from less than .05 inch (1 mm) to blocks as much as 6 feet (2 meters) across. The silt, volcanic ash and blocks were deposited layer on layer as the eruption proceeded. The volcanic rock composed of such a mixture of tiny to very large fragments is known as **tuff breccia** (Figure 19-1). Maars typically are surrounded by a ring of tuff or tuff breccia.

About 650 feet (200 meters) north of the steep trail into the crater is a ridge nose where the tuff breccia is very well exposed. (Hint: it is easier to ascend this ridge nose than to descend it.) Here you can see where some of the big blocks fell with considerable impact, disrupting the strata below. Yet many other blocks did not disrupt the strata; probably these blocks were carried in laterally together with much ash and silt. They were transported by eruption clouds that moved radially out from the vent at very high velocity.

As the trail nears the edge of a steep drop, you walk from tuff breccia onto cinders of the old cinder cone exposed in the cliffs to the south. Note that the trail turns north before the steep drop.

Figure 20-1. Dike of porphyritic basalt. The dark margins of the dike in this picture are cinders that were baked by the heat of magma in the dike and adhered to its sides.

GETTING THERE

Follow the trail north down to a gully and then walk generally downhill toward, but staying safely back from the edge of the cliffs. Walk south about 150 feet (50 m) along the bench to where a wall of dense rock projects out from the cinders almost to the cliff. Descent beyond this point is generally prohibited as of this writing.

DIKE

20

A dike is a tabular mass of rock that cooled from magma injected along a fracture that cuts the layers of surrounding rock. This particular dike is blade-shaped (Figure 20-1). It is unusual in that its base or "keel" is exposed in the cliff and its crest is horizontal and visible high above you. This dike was injected horizontally through the cinders from a source in the southeastern part of the crater. That source was the vent of the old cinder cone. Probably this dike is only slightly younger than the cinders around it.

The cavities .05 - 0.4 inch (1-10 mm) across in the dike represent gas bubbles that were trapped in the cooling magma. Some gas bubbles accumulated to form a large cavity at the top of the dike. Within the dike, the tablets and irregular pieces of clear, glassy material 0.5 - 1 inches (1 or 2 centimeters) across are crystals of plagioclase feldspar. These and crystals of greenish brown olivine grew in the magma before it was erupted and emplaced as a dike. Please do not damage or remove any of the dike.

Figure 21-1. East side of the Sierra Pinacate. Note the canyons cut into the old Santa Clara shield rocks. A young lava flow from Carnegie Peak (highest point on the horizon) is evident in the middle distance.

GETTING THERE

Drive 1.6 miles (2.6 kilometers) east from Crater Elegante; take the right fork, and continue about 9 miles (15 kilometers) to join a road south from Cerro Colorado. Travel 3.9 miles (6.3 kilometers) south on this road and turn off to the right on a road signed "Cono Rojo." Proceed 11.3 miles (18.2 kilometers) up the mountain to the end of this rough road.

ERODED PINACATE VOLCANICS

*T*his road passes over old desert pavement (Feature 6) in several places as it ascends the mountain. Stones in the pavement commonly are heavily coated with rock varnish (Feature 1). Many of these desert surfaces are very old and little disturbed in thousands of years. Rare stone artifacts that have been found embedded in the pavements indicate human occupation of the Pinacate thousands of years ago and perhaps many thousands of years ago. These pavement surfaces record ancient archeological evidence found nowhere else and are very fragile. Do not drive off the road!

Evolution of the land surface in the Pinacate is evident, however, if we look back hundreds of thousands of years. Most striking are the deep notches eroded into the flanks of the old Santa Clara shield volcano (Figure 21-1). Its eroded slopes are covered in most places by younger Pinacate basalt flows. Some of these poured down canyons. In other places they spread more broadly, as did the very young flow clearly visible from the road (and in Figure 21-1) when you are well up on the mountain.

At the end of the road is a cinder cone that has been very deeply dissected by erosion. The framework of the original structure is difficult to discern from this remnant, which is hundreds of thousands of years old. Will Tecolote look like this someday?

From the campground at the end of the road you can climb to the top of Pinacate Peak. The distance is about 4 miles (6 or 7 kilometers) depending on your route. The elevation change is 2560 feet (780 meters). The route is ill-defined in places and commonly rough or on loose cinder. However, the view from the top is superb. Be sure to carry adequate water and take appropriate safety precautions.

SUGGESTED READINGS

Biekman, H.M., Haxel, G.B., and Miller, R.J., 1995, Geologic map of the Tohono O'Odham Indian Reservation, southern Arizona: U.S. Geological Survey Miscellaneous Geologic Investigations Map I-2017, scale 1:125,000, 2 sheets. (Features 3, 5, and 8)

Brown, J.L., 1993, Interpretive geologic map of the Mt. Ajo quadrangle, Organ Pipe National Monument, Arizona: U.S. Geological Survey Open File Report 92-93, Scale 1- 24,000, 13 p. (Features 3 and 5)

Bull, W.B. , 1991, Geomorphic responses to climatic change: Oxford, Oxford University Press, 326 p. (p. 155-157 Features 1 and 8)

Carruth, R.L., 1996, Hydrogeology of the Quitobaquito Springs, and La Abra Plain, Arizona, Organ Pipe Cactus National Monument, and Sonora, Mexico: U.S. Geological Survey Water-Resources Investigations Report 95-4295, 23 p. (Feature 9)

Felger, R.S., Warren, P.L., Anderson, L.S., and Nabhan, G.P., 1992, Vascular plants of a desert oasis: Flora and ethnobotany of Quitobaquito, Organ Pipe Cactus National Monument, Arizona: Proceedings of the San Diego Society of Natural History, no. 8, p. 1-39. (Feature 9)

Gerson, R., 1982, Talus relics in deserts: A key to major climate fluctuations: Israel Journal of Earth Science, v. 31, p. 123-132. (Feature 4).

Gilluly, James, 1946, The Ajo mining district, Arizona: U.S. Geological Survey Professional Paper 209, 112 p. (Feature 4)

Gutmann, J.T., 1976, Geology of Crater Elegante, Sonora, Mexico: Geological Society of America Bulletin, v. 87, p. 1718-1729. (Features 17, 19, and 20)

_____, 1979, Structure and eruptive cycle of cinder cones in the Pinacate Volcanic Field and the controls of Strombolian activity: Journal of Geology, v. 87, p. 448-454. (Features 14 and 20)

Hayden, J.D., 1976, Pre-Altithermal archaeology in the Sierra Pinacate, Sonora, Mexico: American Antiquity, v. 41, p. 274-289. (General reference on the archaeology of the Pinacate Biosphere Reserve)

_____, 1998, The Sierra Pinacate: Tucson, University of Arizona Press, 87 p. (General reference on the Pinacate Biosphere Reserve)

Haxel, G.B., Tosdal, R.M., May, D.J., and Wright, J.E.,1984, Late Cretaceous and early Tertiary orogenesis in south-central Arizona: thrust faulting, regional metamorphism, and granitic plutonism: Geological Society of America Bulletin, v. 95, p.631-653 . (Features 10 and 11)

Hornaday, W.T., 1908, Campfires on desert and lava: New York, Charles Scribners and Sons, 366 p. (General reference on the Pinacate Biosphere Reserve)

Ives, R.L., compiled by Byrkit, J.W., edited by Dahood, K.J., 1989, Land of lava, ash, and sand – The Pinacate region of northwestern Mexico: Tucson, The Arizona Historical Society, 239 p. (General reference on the Pinacate Biosphere Reserve)

Lynch, D.J., 1981, Genesis and geochronology of alkaline volcanism in the Pinacate Volcanic Field of northwestern Sonora, Mexico: Tucson, University of Arizona, Ph.D. dissertation, 248 p. (General reference on the Pinacate Biosphere Reserve)

Lynch, D.J. and Gutmann, J.T., 1987, Volcanic structures and alkaline rocks in the Pinacate Volcanic Field of Sonora, Mexico: Arizona Bureau of Geology and Mineral Technology Special Paper 5, p. 309-322. (General reference on the Pinacate Biosphere Reserve)

Potter, R.M., and Rossman, G.R., 1977, Desert varnish: The importance of clay minerals: Science, v. 196, p. 1446-1448. (Feature 1)

_____, 1979, Mineralogy of manganese dendrites and coatings: American Mineralogist, v. 64, p. 1219-1226. (Feature 1)

_____, 1979, The manganese and iron oxide mineralogy of desert varnish: Chemical Geology, v. 25, p. 79-94. (Feature 1)

Reynolds, S. J., 1988, Geologic map of Arizona: Arizona Geological Survey Map 26, Scale 1:1,000,000. (General reference on Organ Pipe Cactus National Monument)

Smiley, T.L., Nations, J.D., Péwé, T.L. , and Schafer, J.P., editors, 1984, Landscapes of Arizona, the geological Story: Lantham, MD, University Press of America, 505 p. (General geologic reference)

Spencer, J.E., Richard, S.M., Reynolds, S.J., Miller, R.J., Shafiqullah, M., Gilbert, W.G. and Grubensky, M.J., 1995, Spatial and temporal relations between mid-Tertiary magmatism and extension in southwestern Arizona: Journal of Geophysical Research, v. 100, no. B7, p. 10, 321, 351. (Features 3 and 5)

Stratum, I., and Francis, S.C., 1986, The Influence of scree accumulation and weathering on the development of steep mountain slopes, in Abrahams, A.D., 1986 Hillslope processes: Boston, Allen and Unwin, p. 245-267. (Feature 4)

Tosdal, R.M., Haxel, G.B., and Wright, J.E., 1989, Jurassic geology of the Sonoran Desert region, southern Arizona, southeastern California, and northernmost Sonora: Construction of a continental-margin magmatic arc, in Jenney, J.P., and Reynolds, S.J., eds., Geological evolution of Arizona: Arizona Geological Society Digest 17, p. 397-434. (Features 10 and 11)

Tosdal, R.M., Haxel, G.B., Anderson, T.H., Connors, C.D. , May, D.J., and Wright, J.E., 1990, Highlights of Jurassic, Late Cretaceous to Early Tertiary, and Middle Tertiary tectonics, south-central Arizona and north-central Sonora, Gehrels, G.E., and Spencer, J.E., eds., Geologic excursions through the Sonoran Desert region, Arizona and Sonora: Arizona Geological Survey Special Paper 7, p. 76-88. (Features 10 and 11)

Tosdal, R.M., Peterson, D.W., May, D.J., LeVeque, R.A., and Miller, R.J., 1986, Reconnaissance geologic map of the Mount Ajo and part of the Pisinimo Quadrangles, Pima County, Arizona: U.S. Geological Survey Miscellaneous Field Studies Map MF-1820, scale 1:62,500. (Features 3 and 5)

VandenDolder, E.M., 1995, How geologists tell time: Arizona Geological Survey Down-to-Earth 4, 33 p. (General geologic reference)